THE ENGLISH ROSE

LUKE STEPHENSON

History of English Roses

There is nothing more beautiful than Old Roses with their soft colour palette and perfectly formed, fragrant flowers framed on a shapely, natural looking shrub. Modern roses however, tend to be rather more upright, stiffer in appearance, lacking in fragrance and displaying harsher tones. To David Austin these roses were not as pleasing to the eye as the Old Roses.

The English Roses was developed to encompass the beauty and charm of these once popular roses. David wished to capture the qualities of greater colour range and remontant nature of the more modern Hybrid Teas and Floribundas in his varieties.

Initial crosses were made between the best old roses and modern roses of the day. At first this produced non-repeating roses, further work ensued over many years and repeat flowering qualities were introduced into the breeding line and English Roses were born.

As the English Rose continued to improve, disease resistance also became a desired necessity. To this day this quality remains high on the agenda together along with with beauty of flower, fragrance and habit of growth.

A large number of crosses are made each year producing tens of thousands of seedlings, each one being completely unique. From here, over a 10 year period, these seedlings are trialed and selected for their desired characteristics, until finally, only a possible three are deemed good enough to become commercially available.

OLIVIA ROSE AUSTIN

PRINCESS ANNE

CROWN PRINCESS MARGARETA

KEW GARDENS

BENJAMIN BRITTEN

QUEEN OF SWEDEN

THE MAYFLOWER

THE MILL ON THE FLOSS

GERTRUDE JEKYLL

EMILY BRONTË

DESDEMONA

ANNE BOLEYN

GRAHAM THOMAS

ROALD DAHL

TRANQUILLITY

THE POET'S WIFE

THE LADY OF THE LAKE

SKYLARK

DAME JUDI DENCH

LICHFIELD ANGEL

CHARLES DARWIN

LADY EMMA HAMILTON

JAMES L. AUSTIN

THE LADY'S BLUSH

MOLINEUX

GENTLE HERMIONE

THE ANCIENT MARINER

THOMAS A BECKET

THE PILGRIM

JUDE THE OBSCURE

SCARBOROUGH FAIR

THE LADY GARDENER

DARCEY BUSSELL

VANESSA BELL

BUTTERCUP

SUSAN WILLIAMS-ELLIS

ENGLAND'S ROSE

SIR WALTER SCOTT

LADY OF SHALOTT

BATHSHEBA

PRINCESS ALEXANDRA OF KENT

THE LARK ASCENDING

STRAWBERRY HILL

BOSCOBEL

LOCHINVAR

CARIAD

MUNSTEAD WOOD

GRACE

THE GENEROUS GARDENER

HYDE HALL

PORT SUNLIGHT

WILLIAM AND CATHERINE

THE ALBRIGHTON RAMBLER

MALVERN HILLS

JUBILEE CELEBRATION

GOLDEN CELEBRATION

WISLEY 2008

MAID MARION

HARLOW CARR

IMOGEN

CLAIRE AUSTIN

TOTTERING BY GENTLY

MORTIMER SACKLER

WOLLERTON OLD HALL

FIGHTING TEMERAIRE

David Austin Roses

David Austin's interest in breeding roses began in his teens but it remained nothing more than a hobby until his early forties when running his family farm. Alongside farming, David exercised his thoughts and ideas and began to hybridise and produce rose seedlings.

After some success in the 1960's producing such varieties as Constance Spry, Chianti, Canterbury, Dame Prudence and A Shropshire Lass, David founded his own company in 1969. His objective being to further develop and promote his new type of rose, English Roses as he named them.

During the early 1990's he applied the same principles used for the garden rose in breeding the florist's rose, for weddings and special events.

To David beauty and fragrance were key to his work as he believed that fragrance in the florist's rose had been all but bred out.

David Austin Roses, now in their third generation continue his philosophy dedicated to improving the English Rose.

© Luke Stephenson, Stephenson Press.
All right reserved.
No part of this book may be reproduced without the written permission of the publisher or the artist.

Luke Stephenson would like to thank everyone at David Austin Roses for their help and support thoughout this project. Michael Marriott for his expertise and knowledge also Carl Bennett for the foreword and prologue.

First edition published in 2019
by Stephenson Press

Printed in the UK
ISBN 978-0-9574341-2-7